THE POCKET HISTORIAN

A GUIDE TO SMARTPHONE DOCUMENTARY PRODUCTION

DÄNNA WILBERG
TERRY SHEPHERD

RAMIREZ & CLARK PUBLISHERS

Legal Disclaimer & Limitation of Liability

The information contained in this book is provided for general informational and educational purposes only. It is not intended as, and shall not be understood or construed as, professional or legal advice. The recommendations, opinions, and strategies contained herein are based on the authors' experience and research and are not guaranteed to produce any particular results.

The authors and publisher expressly disclaim all warranties, express or implied, regarding the accuracy, completeness, currency, or suitability of the information provided.

By reading this book, you agree that under no circumstances shall the authors or publisher be held liable or responsible for any errors or omissions in this book or for any direct, indirect, consequential, or other damages or losses that may result from the use of, or reliance on, any information contained herein. The reader assumes all responsibility and risk for the use of this information.

Important Note on Documentary Creation

The creation of a documentary film involves complex legal, ethical, and community considerations, including (but not limited to) permissions, talent/location releases, copyright, fair use, privacy rights, and defamation. The guidance in this book is not an exhaustive substitute for professional consultation.

Readers are **strongly admonished to consult with appropriate qualified legal counsel** (such as a media or entertainment lawyer) and relevant community agencies or stakeholders **before** undertaking any documentary project to ensure full compliance with all applicable laws, regulations, and ethical standards.

CONTENTS

INTRODUCTION

The power to create a documentary film, once the exclusive domain of well-funded production houses and broadcast networks, now rests in the palm of your hand. The modern smartphone, a device of ubiquitous connection and convenience, has undergone a quiet revolution, transforming into a pocket-sized studio capable of capturing cinematic, high-resolution images. This technological democratization has shattered the financial and logistical barriers to entry, placing the tools of production within reach of any storyteller with a vision.

A smartphone enjoys several advantages over a camera kit. Spontaneity, You're always ready to shoot. We live in a time where documenting everything from a restaurant dinner plate to a graduation is recorded on a phone. The time factor, Although we will discuss the important impact of proper lighting in this book, the ability to unholster your phone, tap a few settings and grab high quality video in just about any setting out-performs complex camera setups. Accessibility, The small size of a smartphone allows for easier access to tight spaces, creative camera angles and smooth moves that would require more complex solutions with traditional equipment.

However, the accessibility of the technology does not guarantee the quality of the art. A film is not merely a collection of well-exposed images and clear audio; it is a "creative treatment of actuality". At its core, a compelling documentary is an exercise in masterful storytelling. It must engage the audience on both an emotional and intellectual level, guiding them through a narrative constructed from the unscripted chaos of reality. The most sophisticated camera is rendered useless without a deep understanding of narrative structure, cinematic language, and the disciplined workflow that transforms raw footage into a powerful, cohesive story.

This resource is designed to bridge that critical gap between the tool and the craft. It is a simple, yet comprehensive guide for the new generation of mobile documentarians, providing a roadmap that marries the technical potential of the smartphone with the timeless principles of nonfiction filmmaking. We will move beyond the default settings to explore the manual controls that unlock your phone's true cinematic capabilities. We will delve into the primacy of sound, the art of stabilization, and the techniques for painting with light.

More importantly, we will build a foundation in narrative theory, exploring the established documentary modes that define a film's voice and the three-act structure that gives it a compelling shape. From there, we will navigate the entire post-production pipeline—from the meticulous organization of digital assets to the transformative power of the edit, where the story is truly born.

Finally, we will cover the technical specifications

required to deliver a polished, professional film to streaming platforms and the festival circuit.

Whether you are an aspiring filmmaker taking your first steps or an experienced creator adapting to a more nimble workflow, we hope *The Pocket Historian* will become your essential companion; a testament to the idea that the most powerful instrument in filmmaking is not the camera you own, but the story you are determined to tell.

Dänna Wilberg
Terry Shepherd

1

FOUNDATIONS OF MOBILE DOCUMENTARY STORYTELLING

THE CAPACITY TO PRODUCE A DOCUMENTARY FILM NOW resides in the pocket of nearly every creator. Modern smartphones, paired with high-quality wireless audio systems, have democratized the technical means of production, making it possible to create professional films on a budget.[1] However, technology is merely a tool; the true foundation of a compelling documentary lies in its narrative architecture. Before a single frame is recorded, a filmmaker must possess a clear understanding of documentary language, story structure, and the meticulous planning required to transform reality into a powerful cinematic experience. This section establishes that critical groundwork, moving from narrative theory to the practicalities of pre-production.

The Documentary Blueprint

A successful documentary is not an accident. It is the result of a deliberate and structured approach to storytelling. This chapter explores the essential frameworks that guide a filmmaker's vision, from defining the film's narrative voice through established documentary modes to shaping its emotional arc using the classic three-act structure.

1.1 Defining Your Narrative: An Introduction to Documentary Modes

The first step in crafting a documentary is to determine its fundamental approach to reality. Film theorist Bill Nichols identified six primary modes of documentary filmmaking, which serve as a conceptual framework for defining the film's voice and its relationship with the subject and the audience.[2] The choice of mode is not an academic exercise; it is a foundational decision that dictates the entire technical and logistical strategy, especially for a nimble smartphone-based production.

Expository Mode: This is the classic documentary form, often characterized by an authoritative, omniscient "Voice of God" narration that presents a specific argument or explains a topic.[2] Footage in this mode serves to illustrate and provide evidence for the narrative track. This style, popularized by filmmakers like Ken Burns and common in television news and nature documentaries, is ideal for historical, scientific, or biographical subjects where clarity and information delivery are paramount.[3]

Short Definition: "Explain it to me, clearly, confidently, and with receipts."

Examples:

The Civil War (Ken Burns)

Heavy use of authoritative narration, archival footage,

and structured historical argument; viewer is guided toward a clear interpretation.

Planet Earth (BBC)

David Attenborough's narration frames what we're seeing with scientific context and cause-effect explanation.

Cosmos (Sagan/deGrasse Tyson)

Omniscient voiceover presents scientific facts step-by-step, paired with visuals strictly as evidence.

Observational Mode (Cinéma Vérité): This mode adopts a "fly-on-the-wall" perspective, where the filmmaker strives to be an unobtrusive observer of events as they unfold.[3] Pioneered in the 1960s with the advent of more portable camera equipment, this style avoids narration, interviews, and direct intervention, allowing the audience to draw their own conclusions from the captured reality.[6] This approach demands immense patience and readiness from the filmmaker to capture authentic, unscripted moments. A production strategy for an observational documentary must prioritize long battery life (requiring portable power banks) and substantial storage capacity to accommodate long, uninterrupted takes.[9] The gear setup should be as discreet as possible to minimize its influence on the subjects' behavior.

Short Definition: "Don't explain, just watch life unfold."

Examples:

Primary (Robert Drew)

Follows JFK and Hubert Humphrey with no narration or staged setups; viewer interprets events directly.

Grey Gardens (Maysles Brothers)

Pure fly-on-the-wall approach; the filmmakers simply

witness eccentric lives without overt shaping or commentary.

Hoop Dreams (Steve James)

Years of observational shooting capture real struggles, allowing coaches, families, and systems to reveal themselves naturally.

The War Room (Pennebaker)

Access to behind-the-scenes politics with no guiding narration, the campaign speaks for itself.

Participatory Mode: In this mode, the filmmaker becomes an active part of the story, interacting with and interviewing subjects on camera.[3] This interactive style, famously employed by filmmakers like Michael Moore, presents the filmmaker's journey and perspective as central to the narrative.[3] A participatory documentary necessitates a robust audio strategy. For instance, a dual-transmitter wireless microphone system, such as the DJI Mic 2 combo, becomes essential to capture both the subject's and the filmmaker's audio with equal clarity and quality, underscoring the direct link between narrative choice and equipment selection.[10]

Short Definition: "I'm part of this story, watch me engage."

Examples:

Bowling for Columbine (Michael Moore)

Moore actively interviews, confronts, and inserts himself into political and emotional moments.

Super Size Me (Morgan Spurlock)

The filmmaker becomes the test subject and main character, directly shaping what unfolds on camera.

Stories We Tell (Sarah Polley)

Polley investigates her own family history, openly interviewing relatives and acknowledging filmmaking choices.

Louis Theroux documentaries

His interactions, awkward, provocative, charming, drive the narrative, not just the subjects.

Reflexive Mode: This mode turns the camera on the filmmaking process itself, questioning the nature of documentary truth and the relationship between the filmmaker and the audience.[3] By revealing the "man behind the curtain," reflexive documentaries challenge the conventions of the form and provoke viewers to consider how reality is constructed on screen.[4]

Short Definition: "You're watching a documentary, let's talk about that."

Examples:

Man with a Movie Camera (Dziga Vertov)

Constantly reveals filming equipment and editing, making the filmmaking process the subject itself.

F for Fake (Orson Welles)

Welles exposes tricks of editing and storytelling while questioning documentary truth.

Exit Through the Gift Shop (Banksy)

Blurs authenticity and authorial control, making viewers question what is real and what is constructed.

Sherman's March

The film becomes about the filmmaker's failed project and personal distractions, calling out its own premise.

Poetic Mode: Forsaking linear narrative, the poetic mode focuses on creating a specific mood, tone, or visual association.[2] It arranges footage to evoke a subjective, often abstract, interpretation of the subject, prioritizing aesthetic and rhythmic qualities over rhetorical arguments.[2]

Short Definition: "Feel this, don't worry about understanding it."

Koyaanisqatsi

Uses music, montage, and visual rhythm, zero narration, to convey theme through emotion and pattern.

Rain (Joris Ivens)

A lyrical visual study of weather and movement rather than an informational narrative.

The House Is Black

Juxtaposes poetic narration with stark imagery to evoke feeling over analysis.

Meshes of the Afternoon (Maya Deren)**

Dream-logic editing and symbolism create personal interpretation instead of objective meaning.

Performative Mode: This mode emphasizes the subjective experience and emotional response of the filmmaker to the subject matter.[4] It presents a personal, often autobiographical, perspective to explore larger political or historical realities, using the filmmaker's journey as a window into the topic.[8]

Short Definition: "My lived experience *is* the argument."

Examples:

Tongues Untied (Marlon Riggs)

Merges personal narrative, poetry, and activism to express the Black gay experience emotionally and politically.

Waltz with Bashir (Ari Folman)

Animated memories convey trauma, subjective truth matters more than factual visuals.

Tarnation (Jonathan Caouette)

Deeply personal story built from the filmmaker's own archive; raw interiority over traditional documentary form.

Paris Is Burning

Centers marginalized voices with emotional storytelling, subjective identity as political statement.

1.2 Structuring Your Story: Applying the Three-Act Structure

Regardless of the mode, every great documentary tells a story. The most effective and universally understood framework for storytelling is the three-act structure, which organizes real-world events into a compelling narrative with a clear beginning, middle, and end.[12] This structure provides focus during shooting and is an invaluable guide during the editing process.[13]

Act 1: The Setup. The beginning of the film must hook the audience immediately.[15] This act introduces the main characters or subject, establishes the central conflict or question, and presents an "inciting incident"—an event that launches the protagonist on their journey.[13] It sets the tone and poses the dramatic questions that the rest of the film will answer.[16]

Act 2: The Confrontation. This is the main body of the documentary, where the narrative deepens and tension builds.[16] The protagonist encounters escalating obstacles in their attempt to resolve the central conflict. This section is composed of numerous scenes that build upon one another, often revealing a worsening situation that keeps the audience engaged and wondering how the challenges will be overcome.[13]

Act 3: The Resolution. The story culminates in a climax, the moment of peak emotional intensity where the central conflict is confronted directly.[13] This act brings resolution to the questions posed in Act 1, tying up loose ends and providing the audience with a sense of closure, a new understanding, or a call to action.[13]

While the three-act structure is a powerful guide, filmmakers can also employ other narrative strategies, such as a strictly chronological structure, a parallel structure that weaves together multiple related stories, or a non-linear structure that uses flashbacks to create suspense and reveal information in an unconventional way.[17]

1.3 From Concept to Shot List: The Importance of Pre-Production

Success in documentary filmmaking begins with meticulous planning long before the camera starts rolling.[1] This pre-production phase is where the conceptual framework is translated into a practical plan of action.

The first step is to clearly define the documentary's core message and identify the target audience.[1] This clarity of purpose will guide every subsequent decision. From this, a foundational outline or storyboard should be created.[1] A storyboard visually plans out key shots and sequences, while a detailed outline structures the narrative flow.

This planning culminates in a comprehensive shot list. This document ensures that all essential footage is captured, preventing critical gaps in the edit. It should detail the primary interview footage (A-roll) needed to drive the narrative, as well as the supplementary visual material (B-roll or cutaways) required to illustrate the story, cover edits, and add visual interest.[9] A well-conceived shot list is the filmmaker's roadmap in the field, ensuring that production is focused, efficient, and serves the ultimate goal of the story.

2

THE MOBILE PRODUCTION TOOLKIT: CAPTURING CINEMATIC REALITY

Mastering the Smartphone Camera

THE MODERN SMARTPHONE IS A POWERFUL FILMMAKING device, capable of recording in 4K resolution, capturing slow motion, and offering manual control over key cinematic variables.[19] However, unlocking this potential requires moving beyond the default settings and augmenting the phone with essential accessories. This section provides a comprehensive technical guide to the hardware, software, and techniques needed to capture professional-quality video and audio, transforming a consumer device into a veritable pocket studio.

The journey from casual video to cinematic documentary begins with taking full command of the smartphone's camera. This requires bypassing the automated systems designed for convenience and embracing the granular control offered by specialized applications.

2.1 BEYOND AUTO MODE: Unleashing Manual Control

. . .

STOCK CAMERA APPLICATIONS are engineered for point-and-shoot simplicity, constantly making automated adjustments to focus, exposure, and color. While suitable for everyday snapshots, this automation is the enemy of professional filmmaking, as it can ruin a shot by changing settings mid-take.[19]

The first and most critical step is to install a manual camera application. Apps like **Blackmagic Camera** (free) or **FiLmic Pro** are indispensable tools that transform the smartphone into a professional camera, providing direct control over the core elements of cinematography.[19] These apps unlock the ability to manually set and lock focus, exposure, frame rate, shutter speed, ISO, and white balance.[19]

A fundamental technique enabled by these apps is the **Focus & Exposure Lock**. By tapping and holding on a specific point on the screen, the filmmaker can lock the camera's focus and exposure settings.[9] This prevents the phone from "hunting" for focus or adjusting the brightness if the subject moves or lighting conditions shift slightly, ensuring consistency throughout a take—a hallmark of professional footage.[18]

2.2 The Exposure Triangle on a Smartphone

MANUAL CAMERA APPS grant access to the three pillars of photographic exposure, allowing the filmmaker to balance them for both technical accuracy and creative effect.[20]

ISO: This setting determines the camera sensor's sensitivity to light. A low ISO value (e.g., 100, 200) produces the cleanest, sharpest image with the least amount of digital noise, but it requires ample light.[20] As the ISO is increased (e.g., 1600, 3200), the sensor becomes more sensitive, allowing for shooting in darker conditions. However, this comes at the cost of introducing digital noise or "grain," which can degrade image quality.[20] The goal is always to keep the ISO as low as possible for the given lighting conditions.[23]

Shutter Speed: This value dictates the duration for which the sensor is exposed to light for each frame of video.[20] Shutter speed has a direct impact on motion blur. To achieve a natural, cinematic motion blur that mimics what the human eye perceives, filmmakers adhere to the "180-degree shutter rule," which states that the shutter speed should be double the frame rate.[20] For example, when shooting at 24 frames per second (fps), the ideal shutter speed is 1/48th of a second (or the closest available setting, typically 1/50th).[20] A faster shutter speed (e.g., 1/1000s) will freeze motion, resulting in a crisp but potentially stuttery look, while a slower shutter speed will create more pronounced motion blur.[20]

Aperture: On dedicated cameras, the aperture is a physical iris in the lens that controls the amount of light entering and affects the depth of field. Most smartphone lenses have a fixed physical aperture.[22] However, some phones and apps use software to simulate a shallow depth of field, often called "Cinematic Mode," which can be used to create a blurry background and isolate the subject.[20]

White Balance (WB): This setting controls the color temperature of the image, ensuring that white objects appear white regardless of the lighting source (e.g., warm indoor light, cool daylight).[22] While auto white balance is convenient, it can shift during a take if lighting changes, causing distracting color fluctuations. Manually setting the white balance to a specific Kelvin value (e.g., 5600K for daylight) or using a preset ensures consistent, accurate color throughout a scene.[20]

2.3 Resolution and Frame Rates: Technical Choices, Creative Impact

The technical settings for resolution and frame rate are not merely about image quality; they are fundamental creative choices that impact the entire production and post-production workflow.

Resolution (4K vs. 1080p): It is highly recommended to shoot in the highest resolution available, typically 4K (3840×2160 pixels).[1] Even if the final documentary will be delivered in 1080p (HD), shooting in 4K provides significant advantages in post-production. The larger frame allows the editor to crop, reframe, or "punch in" on a shot to create a close-up from a wider take without losing image quality. It also provides extra pixels for digital stabilization.[1] The main trade-off is that 4K files consume a large amount of storage space, so adequate preparation is necessary.[9]

Frame Rate (fps): The choice of frame rate has a profound effect on the aesthetic feel of the documentary.

24fps: This is the long-standing frame rate of cinema and is the standard for achieving a "film look." It produces a natural amount of motion blur that audiences associate with professional movies.[1]

30fps: This is the standard for traditional broadcast television and is common for online videos. It results in a slightly smoother, more "video-like" motion compared to 24fps.[18]

60fps / 120fps: These higher frame rates are used specifically for capturing slow-motion footage. When a clip shot at 60fps is played back on a 24fps timeline, it will play at 40% speed, creating smooth and dramatic slow motion. This can be a powerful tool for adding emphasis to a key action or emotional moment.[1] It is critical to maintain a consistent primary frame rate (e.g., 24fps) throughout the project and only use higher frame rates for shots intended for slow motion, as mixing frame rates can lead to playback issues.[19]

Scenarios and Settings:

Standard Interview (Daylight)

Resolution: 4K

Frame Rate: (fps) 24

Shutter Speed: 1/48 or 1/50

ISO Range: 100-400

White Balance (WB): 5600K (Daylight)

Lock AE/AF on subject. Use gridlines for Rule of Thirds framing.[19]

Cinematic B-Roll (Daylight)

Resolution: 4K

Frame Rate: (fps) 24

Shutter Speed: 1/48 or 1/50

ISO Range: 100-400

White Balance (WB): 5600K (Daylight)

Use a gimbal for smooth motion. Avoid digital zoom; move closer instead.[9]

Dramatic Slow Motion

Resolution: 4K

Frame Rate (fps): 60 or 120

Shutter Speed: 1/120 or 1/240

ISO Range: As low as possible

White Balance (WB): Manual or Auto

High frame rates require significantly more light to achieve proper exposure.

Interior Interview (Low Light)

Resolution: 4K

Frame Rate (fps): 24

Shutter Speed: 1/48 or 1/50

ISO Range: 400-1600

White Balance (WB): Manual (match lights)

Add external lighting (e.g., three-point setup) to keep ISO low and avoid noise.[9]

3

THE PRIMACY OF SOUND: WIRELESS AUDIO WITH DJI MIC

VISUALS MAY DRAW AN AUDIENCE IN, BUT IT IS THE QUALITY OF the audio that keeps them engaged. Poor sound is the most common and jarring flaw in amateur productions, and it is a mistake that is easily avoidable.[19] The built-in microphone on a smartphone is designed for phone calls, not for capturing the nuanced, clear dialogue required for a documentary. It is an omnidirectional microphone that is invariably too far from the subject, resulting in echoey, distant audio contaminated by ambient noise.[9] For this reason, an external microphone is the single most important accessory a mobile documentarian can invest in.[19]

3.1 Why External Audio is Non-Negotiable

CLEAN, intelligible dialogue is the backbone of most documentaries. Whether conducting an interview or capturing a conversation, the goal is to isolate the subject's voice from the surrounding environment. A wireless lavaliere microphone system, such as the DJI Mic, accomplishes this by placing a small microphone directly on the subject, ensuring a close, consistent audio source regardless of the camera's position.[18] This immediately elevates the production value from amateur to professional.

3.2 Technical Setup: Connecting DJI Mic 2 to Smartphones

Modern wireless systems are designed for seamless integration with smartphones. The DJI Mic 2, for example, offers several connection methods:

Direct Receiver Connection: The most robust and reliable method involves plugging the DJI Mic receiver directly into the smartphone's charging port. The system comes with interchangeable USB-C and Lightning adapters, making it compatible with the vast majority of modern Android and iOS devices.[10] The phone automatically recognizes the receiver as an external audio input.[29]

Bluetooth Connection: The DJI Mic 2 transmitter also features the ability to connect directly to a smartphone via Bluetooth.[26] This is a convenient option that frees up the phone's charging port for a power bank, but it may introduce a small amount of latency (delay) that needs to be corrected in post-production.

Pairing and Charging: For kits that include a charging case, the transmitters (TX) and receiver (RX) are typically pre-linked and will pair automatically when removed from the case.[26] The case also serves to charge all components simultaneously.[26] If components are purchased separately, a simple manual linking process is required.[26] Before any shoot, it is essential to ensure all components are fully charged.[29]

3.3 Recording Modes Deep Dive: Mono, Stereo, and the Safety Track

The DJI Mic system provides several recording modes, each tailored to specific documentary scenarios.[26] Understanding these modes is crucial for capturing the most useful audio for the edit.

Mono (M): In this mode, the audio from one or two transmitters is combined into a single audio channel. This is the simplest option, suitable for a single-person interview or when two subjects are speaking and their audio does not need to be mixed independently.

Stereo (S): This mode records each transmitter to a separate audio channel (left and right). This is the essential mode for any two-person interview. By keeping the audio streams separate, the editor has complete control in post-production to adjust the volume level of each person independently, balance the conversation, and remove noise from one channel without affecting the other.

Mono with Safety Track (Ms): This is a critical feature for unpredictable recording environments. In this mode, the system records the main mono audio track as well as a duplicate backup track at a lower volume, typically -6dB.[26] This "safety track" is an insurance policy against audio clipping—the harsh distortion that occurs when a sound is too loud for the microphone's input level. If an interviewee suddenly laughs loudly or a door slams, the main track might distort, but the -6dB safety track will remain clean and usable, saving the take.

The DJI Mic 2's internal recording capability introduces a feature that fundamentally alters the workflow for solo documentary shooters: 32-bit float recording.[30] In traditional 16-bit or 24-bit audio recording, the filmmaker must carefully set the gain (input level) to ensure the signal is strong enough without being so loud that it clips. This is a delicate balancing act that is difficult to manage for a solo operator who is simultaneously framing the shot, managing lighting, and engaging with the subject. 32-bit float audio has such an immense dynamic range that it is virtually impossible to clip the recording at the source. If the audio signal is too loud and appears distorted, the waveform can be simply lowered in post-production software, fully restoring the clean, undistorted audio. This technology effectively shifts the high-stakes task of "getting levels right" from the unpredictable field environment to the controlled post-production suite, dramatically increasing the reliability of capturing perfect audio for a one-person crew.

3.4 Mic Placement and Troubleshooting

Proper microphone placement is key to capturing clear audio. The lavaliere transmitter should be clipped to the subject's collar, lapel, or the edge of their shirt, approximately 6-8 inches from their mouth.[29] The included clip magnet offers more flexible placement options on different types of clothing.[10] When shooting outdoors, even in a light breeze, the included windscreen (often called a "dead cat") is essential for reducing wind noise.[10]

One of the greatest challenges in smartphone filmmaking is **audio monitoring.** It is vital to confirm that clean audio is being recorded.

Visual Monitoring: Use a camera app like Blackmagic Camera that displays on-screen audio meters (VU meters). These meters provide a visual representation of the audio levels, allowing the filmmaker to see if the signal is too low (quiet) or too high (clipping).[31]

Aural Monitoring: To listen to the audio in real-time, a dedicated audio interface with a headphone jack is the best solution.[31] This small device plugs into the smartphone's port and allows both a microphone and headphones to be connected simultaneously. Some wireless receivers also have a 3.5mm headphone output for direct monitoring.

Troubleshooting Wireless Interference: Wireless systems operate on radio frequencies (RF) and are susceptible to interference, which can manifest as static, popping sounds, or complete signal dropouts.[32]

Common Causes: Interference can be caused by other wireless devices (Wi-Fi routers, other microphones), strong TV or radio broadcast signals, and physical obstructions like walls, metal objects, or even the human body blocking the signal.[32]

Solutions:

Maintain Line of Sight: Ensure there is a clear, unobstructed path between the transmitter (on the subject) and the receiver (on the phone).[32]

Minimize Distance: Keep the distance between the transmitter and receiver as short as is practical.[36]

Change Frequency: If interference occurs, most wireless

systems allow the user to switch to a different frequency or channel to find a clear signal.[33]

Check Batteries: Low battery power in the transmitter can lead to a weak and unreliable signal. Always start a shoot with fresh, fully charged batteries.[32]

4

ACHIEVING FLUID MOTION: STABILIZATION WITH GIMBALS

Unstable, shaky footage is one of the most immediate indicators of an amateur production.[19] While modern smartphones incorporate sophisticated internal stabilization systems, achieving truly smooth, cinematic motion requires an external mechanical stabilizer, known as a gimbal.

4.1 Mechanical vs. Digital Stabilization

Smartphones employ two primary types of built-in stabilization:

Optical Image Stabilization (OIS): A physical system where the lens or sensor assembly is moved by tiny motors to counteract small shakes and jitters.

Electronic Image Stabilization (EIS): A software-based system that crops into the image and shifts the frame to smooth out movement.

While these systems are effective for casual video, they

are insufficient for the dynamic movements often required in documentary filmmaking, such as walking, tracking a subject, or executing smooth pans.[37] Furthermore, software-based stabilization can introduce undesirable artifacts like warping and always results in a reduction of image quality due to the necessary cropping of the frame.[37]

A **3-axis mechanical gimbal** is the superior solution. This handheld device uses a system of brushless motors and sensors to actively counteract the filmmaker's movements across the pan (left-right), tilt (up-down), and roll axes.[38] The result is exceptionally smooth, "buttery" footage that appears to float, mimicking the look of professional Steadicam rigs.[38]

4.2 Operating a 3-Axis Gimbal (e.g., **DJI Osmo Mobile, Insta3_S60 Flow)**

Effective gimbal operation is a skill that combines proper setup with practiced physical technique.

Balancing: Before powering on the gimbal, the smartphone must be correctly mounted and balanced in the cradle. An unbalanced phone forces the motors to work harder, which drains the battery and can result in jittery footage or motor failure.

Basic Movements: Even with a gimbal, the operator's movement influences the final shot. To minimize vertical bounce when walking, operators should practice the "ninja walk"—walking heel-to-toe with knees bent to absorb impact. The gimbal's built-in joystick allows for precise and smooth pans and tilts that are impossible to achieve by hand.

Advanced Techniques: A gimbal unlocks a range of cinematic shots that add significant production value. These include dynamic tracking shots that follow a moving subject, parallax movements where the camera moves laterally to create a sense of depth as foreground elements pass by faster than the background, and dramatic low-angle shots that would be shaky and unstable if attempted handheld.[37]

4.3 Leveraging Intelligent Features

Modern gimbals from manufacturers like DJI, Insta360, and Zhiyun are not just stabilizers; they are intelligent filmmaking tools that integrate with powerful companion apps.[37]

ActiveTrack (Subject Tracking): This is a transformative feature for solo documentarians. By drawing a box around a subject on the smartphone screen, the gimbal's AI will lock onto that subject and automatically pan and tilt to keep them perfectly framed as they move.[38] This allows a single operator to capture complex following shots or even film themselves while moving, effectively acting as an automated camera operator.[37]

Automated Movements: Gimbal apps often include preprogrammed shooting modes that automate complex cinematic techniques. These can include motion-lapses (a timelapse with added camera movement), hyper-lapses (a stabilized, moving time-lapse), and sweeping panoramic shots, all of which can add a highly polished and professional feel to a documentary's B-roll footage.[38]

5

PAINTING WITH LIGHT

LIGHT IS THE FUNDAMENTAL MEDIUM OF CINEMATOGRAPHY. For smartphone cameras, with their relatively small image sensors, an abundance of good-quality light is not just a creative choice—it is a technical necessity for producing a clean, professional-looking image.[21] Mastering lighting does not require expensive equipment; it begins with understanding how to observe and manipulate the light that is already available.[1]

5.1 HARNESSING NATURAL LIGHT: The Cinematographer's Best Tool

NATURAL LIGHT IS OFTEN the most beautiful and flattering light source available to a documentarian, and learning to use it effectively is a crucial skill.

Window Light: When filming an interview indoors, the single best source of light is a large window. Position the subject so they are facing the window, with the camera placed between them and the window.[21] The window acts as a large, soft, and diffuse light source that wraps gently around the subject's face, minimizing harsh shadows and creating a pleasing look.[21]

Golden Hour: The period shortly after sunrise and before sunset is known as the "golden hour." During this time, the sun is low in the sky, and its light is soft, warm, and directional, creating long shadows and a cinematic quality.[1] This is the ideal time for shooting outdoor scenes and B-roll. Conversely, filmmakers should avoid shooting in harsh, direct midday sun, which creates unflattering, hard shadows on the face.[21] If shooting midday is unavoidable, finding a shaded area can provide softer, more even lighting.[21]

Avoiding Backlight: A common mistake is to place the subject with a bright light source, such as a window or the sun, directly behind them.[25] The smartphone's auto-exposure system will try to expose for the bright background, plunging the subject into a dark, featureless silhouette. The filmmaker must always be aware of the direction of the primary light source and position the subject accordingly.[25]

5.2 The Professional Interview Setup: Three-Point Lighting on a Budget

For controlled interview settings, the industry-standard technique is **three-point lighting.** This setup uses three distinct light sources to shape the subject, control shadows, and create a sense of depth that visually separates them from the background.[41] This professional look can be achieved affordably using inexpensive LED panels, or even creatively with household lamps and reflectors.

Key Light: This is the primary and brightest light source. It is typically placed at a 45-degree angle to the camera and slightly above the subject's eye level.[41] The key light is responsible for illuminating the subject and creating defining shadows, which give their face shape and dimension.[42]

Fill Light: Placed on the opposite side of the camera from the key light, the fill light is a softer, less intense light source.[42] Its purpose is to "fill in" the dark shadows created by the key light, thereby controlling the overall contrast of the scene.[41] The fill light should be about half the brightness of the key light.[41] A simple and effective budget solution is to use a reflector, such as a white foam core board, to bounce some of the key light back onto the shadow side of the subject's face instead of using a second lamp.[1]

Back Light (or Hair Light): This light is placed behind the subject, often out of frame and pointing towards the back of their head and shoulders.[41] It creates a subtle rim of light that outlines their form, visually separating them from the background and adding a professional touch of depth and polish.[42] This final element prevents the subject from looking "pasted" onto the background and makes the image feel more three-dimensional.

Part III: The Digital Darkroom: Post-Production Workflow

The production phase captures the raw materials of the documentary, but it is in post-production where the story is truly forged. This is a meticulous process of organization, narrative construction, and technical refinement that transforms hours of disparate footage into a cohesive and polished film. A disciplined and structured post-production workflow is arguably the most critical factor in achieving a professional result, regardless of the specific editing software used.

6

THE DIGITAL DARKROOM: POST-PRODUCTION WORKFLOW

Building the Foundation: Asset Management and Organization

Documentary projects generate a vast amount of digital material. Without a robust system for organizing these assets from the very beginning, the editing process can quickly descend into a chaotic and inefficient search for files.[15] A logical and consistent folder structure is not an optional step; it is a professional requirement.

6.1 The Professional Folder Structure

A standardized folder structure should be created for every new project. This ensures that every type of asset has a designated home, making files easy to locate for the editor and any collaborators.[47] The following is an industry-standard template that can be adapted for any documentary project:

[Project_Name] (Master Project Folder)

01_PROJECT_FILES: This folder houses the project files created by the non-linear editing (NLE) software (e.g., Adobe Premiere Pro, DaVinci Resolve).[48]

02_FOOTAGE: Contains all the original, untouched raw video files from the smartphone(s).[48]

Within this folder, create sub-folders organized by shooting date and camera/card identifier. This is crucial for maintaining a clear link to the original media. Example: 2024-10-26_iPhone15_Card01.[45]

03_AUDIO: Contains all audio files recorded externally, such as from the DJI Mic system.[48]

The sub-folder structure here should mirror the FOOTAGE folder to make syncing audio and video straightforward. Example: 2024-10-26_DJIMic_Card01.

04_GRAPHICS: A repository for all graphic elements, including titles, lower thirds, logos, and any animations.[48]

05_MUSIC_SFX: Contains all licensed music tracks and sound effects used in the film.[48]

06_EXPORTS: The destination for all rendered versions of the documentary, from rough cuts to the final master file.[49]

07_DOCUMENTS: A central location for all project-related paperwork, such as scripts, call sheets, release forms, and, most importantly, interview transcripts.[48]

6.2 Ingest and Backup Strategy: The 3-2-1 Rule

The first action after any shoot day should be to immediately upload and back up all newly created media.[45] Digital assets are fragile, and a single hard drive failure can wipe out irreplaceable work. The professional standard for data protection is the

3-2-1 Rule: maintain at least **3** total copies of your data, on **2** different types of storage media, with **1** of those copies located off-site.

A practical workflow for a mobile documentarian would be:

1. Copy all footage and audio to a primary external hard drive for editing.
2. Create a second, identical copy on a separate external hard drive that is stored in a safe place.
3. Upload a third copy to a cloud storage service (e.g., Google Drive, Dropbox, Backblaze). This off-site copy protects against local disasters like fire, flood, or theft.

6.3 File Naming Conventions and Metadata

Consistent file naming is essential for an organized workflow, especially when dealing with cameras that may reset file names like MVI_0001 on each new memory card.[53] Renaming files upon ingest using a consistent convention prevents confusion and makes media management far easier. A useful naming scheme could be:

___. For example: 20241026_S01_SH05_Interview-CU_JohnDoe.mov.[55]

. . .

Within the editing software, **metadata** should be used to tag clips with descriptive keywords. This creates a searchable database of all footage. Clips can be tagged with character names, locations, themes ("climate change," "family history"), or shot types ("B-roll," "archival," "drone").[53] This allows the editor to instantly find every shot related to a specific topic, dramatically speeding up the creative process.[57]

7

FROM SPOKEN WORD TO STORY: TRANSCRIPTION AND THE PAPER EDIT

FOR INTERVIEW-DRIVEN DOCUMENTARIES, THE SPOKEN WORD IS the narrative bedrock. The process of converting this spoken word into text—transcription—is a transformative step that unlocks a more efficient and intellectually rigorous way of building the story.

7.1 The Transcription Process

TRANSCRIBING every word of every interview is a foundational step in the documentary editing workflow.[54] A transcript provides a searchable, scannable document of all dialogue, saving the editor from the tedious task of scrubbing through hours of footage to find a specific quote.[54]

Transcription Services: While manual transcription is possible, it is incredibly time-consuming, with an average ratio of four hours of work for every one hour of footage.[58] AI-powered transcription services like Rev.com, Trint, Descript, and Simon Says are highly recommended. They offer fast turnaround times, high accuracy (often >99% for human-verified services), and automatically include crucial features like speaker labels and timecodes.[58] Many modern NLEs, including Adobe Premiere Pro and DaVinci Resolve, now have powerful, built-in transcription features as well.[65]

Verbatim vs. Clean Verbatim: When ordering transcripts, filmmakers can choose between two styles. "Clean Verbatim" removes filler words ("ums," "ahs") and stutters, creating a more readable document.[60] "True Verbatim," however, captures every utterance exactly as it was spoken. For documentary work, True Verbatim is often preferred, as these non-verbal cues can reveal a subject's emotional state, hesitation, or conviction, which can be just as important as the words themselves.[66]

7.2 Constructing the Narrative on Paper (The "Paper Edit")

With transcripts in hand, the filmmaker can begin the **paper edit.** This is a pre-editing process where the narrative structure of the film is built by highlighting, cutting, and pasting sections of the text transcripts into a new master document.[46] This allows the director or editor to construct scenes, arrange arguments, and perfect the story's flow without ever opening the editing software.[70]

This text-based approach is profoundly efficient. It allows for rapid iteration and experimentation with story structure, as rearranging paragraphs of text is far faster than moving video clips on a timeline.[70] Major story problems can be identified and solved at this stage, saving countless hours of work in the edit suite.[54]

The traditional concept of the "paper edit" has undergone a significant digital evolution. While it once involved physical paper or a simple word processor, modern tools have integrated this process directly with the video footage. Platforms like **Descript** and **Simon Says** link the transcript to the corresponding video file in a text-based interface.[61] In this environment, editing the text *is* editing the video.

Deleting a sentence in the transcript automatically removes that segment from the video clip. Rearranging paragraphs reorders the video clips on a timeline. This paradigm shift blurs the line between writing and editing. It empowers story-focused creators, such as directors and producers, to create a timed, viewable assembly cut of the film directly from the transcript, without needing advanced technical skills in a complex NLE. This accelerates the feedback loop and democratizes the initial stages of the editing process, allowing the story to be refined with incredible speed and flexibility.

8

THE ART OF THE EDIT

The edit is where the raw materials—footage, audio, transcripts, and the paper edit—are synthesized into a living, breathing film. This creative and technical process involves choosing the right software tool and progressing through distinct stages of refinement to shape the final narrative.

8.1 Choosing Your Non-Linear Editor (NLE)

The choice of NLE is a significant one, as it will be the primary tool for the duration of the post-production process. The decision should be based on factors like budget, experience level, computer platform, and desired workflow.

DaVinci Resolve: Originally a high-end color grading suite, DaVinci Resolve has evolved into a fully-featured, all-in-one post-production powerhouse. The standard version is completely free and includes incredibly powerful modules for editing, color correction, audio post-production (Fairlight), and visual effects (Fusion).[71] Its professional-grade toolset and lack of cost make it arguably the best value for an independent documentarian. However, its comprehensive nature comes with a steeper learning curve compared to other options.[71]

Adobe Premiere Pro: For many years, Premiere Pro has been the industry standard for a wide range of productions, from YouTube videos to feature films.[72] It is known for its robust editing tools, intuitive interface, and seamless integration with other Adobe Creative Cloud applications like After Effects and Photoshop.[73] Premiere Pro is available via a monthly or annual subscription, which can be a significant ongoing cost.[71]

LumaFusion: For filmmakers committed to a fully mobile workflow, LumaFusion is the undisputed professional-grade NLE for iOS and iPadOS.[19] It offers a remarkable range of features, including multi-track editing, color correction tools, and advanced audio mixing, all within a touch-based interface. It is a one-time purchase, making it a cost-effective option for those editing on an iPhone or iPad.[19]

Free Options for Beginners: Applications like CapCut (cross-platform) and Apple's iMovie (iOS/macOS) are excellent for beginners and short-form content. They are user-friendly and offer a surprising number of features.[9] However, they lack the advanced organizational tools, precise controls, and professional export options required for a feature-length documentary project.

Comparing Some Options

Cost

DaVinci Resolve: Free (Studio version is a one-time fee)

Adobe Premiere Pro: Monthly/Annual Subscription

LumaFusion (iOS): One-time Purchase

Platform

DaVinci Resolve: Windows/Mac/Linux

Adobe Premiere Pro: Windows/Mac

LumaFusion (iOS): iOS/iPadOS

Key Strengths

DaVinci Resolve: All-in-one solution (Edit, Color, Audio, VFX), Powerful free version

Adobe Premiere Pro: Industry Standard, Adobe Creative Cloud Integration

LumaFusion (iOS): Best-in-class mobile-native workflow, professional features

Learning Curve

DaVinci Resolve: High

Adobe Premiere Pro: Medium

LumaFusion (iOS): Low-Medium

Smartphone Workflow

DaVinci Resolve: Desktop-centric; requires transferring files from phone to computer.

Adobe Premiere Pro: Desktop-centric; some integration with Premiere Rush but primarily requires file transfer.

LumaFusion (iOS): Fully mobile native; edit footage directly on the device it was shot on.

8.2 The Editing Stages Demystified

. . .

THE EDITING PROCESS is not a single action but a gradual refinement through several distinct stages. This structured approach helps manage the complexity of a large project and ensures a polished final product.[75]

Assembly Cut: This is the very first pass of the edit. The editor takes the paper edit as a guide and lays all the selected interview clips, scenes, and audio onto the timeline in the correct sequence.[78] The assembly cut is typically very long—often more than twice the final runtime—and extremely rough. Its sole purpose is to get all the narrative building blocks in place.[77]

Rough Cut: In this stage, the assembly is refined into something that begins to resemble a film. The editor trims excess footage, adjusts the pacing of scenes, and removes obvious mistakes.[76] The story structure is solidified, and the overall flow of the narrative becomes clear. The rough cut is the ideal version to share with trusted collaborators or producers for initial feedback, as major structural changes are still relatively easy to make.[75]

Fine Cut: This stage is about polishing and detail. The editor tightens the pacing, perfects the timing of cuts, and smooths out transitions.[75] Placeholder graphics, temporary music, and basic sound design are often added to give a better sense of the final film's emotional tone and rhythm.[76]

Picture Lock: This is a critical milestone. Picture lock means that no further changes will be made to the timing or sequence of the video edit. The story, structure, and pacing are now finalized. Once picture lock is declared, the project is ready to be handed off for the final finishing processes: professional color grading and sound mixing.

8.3 Pacing, Rhythm, and Weaving B-Roll

. . .

BEYOND THE MECHANICAL assembly of clips, the art of editing lies in controlling the film's pacing and rhythm. This is the manipulation of time to control the flow of information and guide the audience's emotional experience.

The strategic use of B-roll (cutaway shots) is a key component of this process. B-roll should do more than simply cover a jump cut in an interview where a sentence has been removed.[59] Effective B-roll adds a new layer of meaning to the film. It can provide visual context for what a subject is describing, create an emotional counterpoint to their words, or introduce a visual metaphor that deepens the film's theme.[9] The interplay between the spoken word (A-roll) and the illustrative visuals (B-roll) is where much of a documentary's narrative power is generated.

9

FINISHING: COLOR, SOUND, AND POLISH

WITH THE PICTURE LOCKED, THE DOCUMENTARY ENTERS THE final finishing stage. This is where the technical and aesthetic polish is applied, elevating the project to a professional standard through meticulous color grading and sound mixing.

9.1 Color Correction vs. Color Grading

THOUGH OFTEN USED INTERCHANGEABLY, color correction and color grading are two distinct processes that are crucial for achieving a cinematic look, especially with smartphone footage, which can often appear overly saturated or digitally sharpened out of the camera.[81]

Color Correction: This is the technical, corrective stage. The goal is to ensure that the footage is clean, consistent, and natural. This involves several steps:

Balancing Exposure: Adjusting the brightness and

contrast so that nothing is too dark (crushed blacks) or too bright (clipped whites).

Setting White Balance: Correcting any color casts so that whites are pure white and skin tones look natural.[83]

Shot Matching: Ensuring that all shots within a single scene have a consistent look in terms of brightness and color, even if they were shot at slightly different times or angles.[83]

Color Grading: This is the creative, stylistic stage. Once the footage is corrected and balanced, the colorist can create a specific "look" or mood that enhances the story's emotional tone.[84] This could involve shifting the colors towards a cool, blue palette for a somber scene, or a warm, golden look for a nostalgic one. LUTs (Look-Up Tables) are often used as a starting point for creating these cinematic looks.[81]

These processes can be performed within the powerful color tools of desktop NLEs like DaVinci Resolve or Adobe Premiere Pro's Lumetri Color panel, or for simpler projects, on mobile apps like VN Video Editor, CapCut, or even Adobe Lightroom Mobile.[85]

9.2 An Introduction to Sound Mixing

Sound mixing, or sound design, is the process of balancing all the audio elements of the film to create a clear, rich, and immersive soundscape.[84] A good sound mix is often unnoticeable, while a bad one can be incredibly distracting. The key components include:

Dialogue Cleanup: The first priority is ensuring the dialogue is clean and intelligible. This involves using audio tools to reduce background noise, remove hums or hisses, and enhance vocal clarity.

Leveling: All audio elements must be balanced. Dialogue should be at a consistent, audible level throughout the film. Music should support the scene emotionally without overpowering the speech. Sound effects should be placed appropriately in the mix to feel natural.[74]

Adding Music & Sound Effects (SFX): The musical score is a powerful tool for guiding the audience's emotions.[59] Sound effects are equally important for creating a believable world. This includes adding ambient sounds (e.g., the sound of a city street, wind in the trees) to make locations feel real, and specific sounds (foley) that might not have been captured cleanly during production (e.g., foot-steps, a door closing).[59]

10

DELIVERY AND DISTRIBUTION

THE FINAL STAGE OF THE DOCUMENTARY FILMMAKING PROCESS involves exporting the finished film into the correct technical formats for its intended destinations. Whether the goal is to share the film on online streaming platforms or submit it to the competitive film festival circuit, understanding the technical specifications for delivery is a critical final step.

Exporting Your Masterpiece

After countless hours of shooting, organizing, and editing, the export process renders the final film into a single, playable file. Choosing the correct settings is vital to ensure the film is viewed at the highest possible quality while being compatible with the target platform.

10.1 Understanding Codecs and Containers

. . .

TWO KEY TERMS govern the world of digital video files: codec and container.

Container: This is the file format or "wrapper" that holds all the different elements of the film together—the compressed video stream, the audio stream, and metadata like subtitles. Common container formats include .MP4 and .MOV.[89]

Codec (Coder-Decoder): This is the algorithm used to compress the massive raw video data into a manageable file size and then decompress it for playback.[91] The choice of codec is the primary factor determining the balance between file size and visual quality.

H.264 (AVC, Advanced Video Coding): This is the most ubiquitous and widely supported video codec in the world.[89] It offers an excellent balance of quality and compression and is the universal standard for web streaming on platforms like YouTube and Vimeo. For maximum compatibility, H.264 is the safest and most recommended choice for online distribution.[90]

H.265 (HEVC, High Efficiency Video Coding): The successor to H.264, this codec is roughly 50% more efficient, meaning it can deliver the same visual quality at half the file size, or higher quality at the same file size.[90] It is ideal for streaming 4K and HDR content, but its adoption is not yet as universal as H.264, and some older devices may not support it.[90]

Apple ProRes: This is a family of professional, high-quality codecs designed for use during the editing process, not for final web delivery. ProRes files preserve a very high level of image detail but are extremely large.[94] A ProRes file is often used as the final "master" file from which all other compressed delivery versions are made.

. . .

10.2 Optimal Export Settings for Streaming Platforms

When exporting for online platforms like YouTube and Vimeo, the goal is to create a file that is high quality but also compressed efficiently for smooth streaming. The following settings, primarily using the H.264 codec within an MP4 container, are recommended.[96]

Video Settings:

Format/Container: MP4 [97]

Codec: H.264 [97]

Resolution: Match the project's resolution (e.g., 1920x1080 for HD, 3840x2160 for 4K).[97]

Frame Rate: Match the source frame rate of the project (e.g., 24fps).[96]

Bitrate Encoding: VBR (Variable Bitrate), 1-pass or 2-pass. VBR is more efficient than CBR (Constant Bitrate) as it allocates more data to complex scenes and less to simple ones.[96]

Target Bitrate: This determines the file size and quality. For 1080p HD, a target bitrate of 12-16 Mbps is a good starting point. For 4K, a target of 40-60 Mbps is recommended.[97]

Audio Settings:

Format/Codec: AAC (Advanced Audio Coding) is the standard for web video.[96]

Sample Rate: 48 kHz.[96]

Channels: Stereo.[96]

Bitrate: 320 kbps for high-quality audio.[98]

For archival and mastering purposes, uncompressed audio formats like **WAV** or **AIFF** at 24-bit depth and 48 kHz sample rate should always be used to preserve the maximum possible audio fidelity.[99]

Settings Suggestions

Container

YouTube/Vimeo (1080p HD): .MP4

YouTube/Vimeo (4K UHD): .MP4

High-Quality Archival Master: .MOV

Video Codec

YouTube/Vimeo (1080p HD): H.264

YouTube/Vimeo (4K UHD): H.264 (or H.265 for smaller files)

High-Quality Archival Master: Apple ProRes 422 HQ (or higher)

Resolution

YouTube/Vimeo (1080p HD): 1920x1080

YouTube/Vimeo (4K UHD): 3840x2160

High-Quality Archival Master: Source Resolution (e.g., 3840x2160)

Bitrate (VBR)

YouTube/Vimeo (1080p HD): 12-16 Mbps

YouTube/Vimeo (4K UHD): 40-60 Mbps

High-Quality Archival Master: N/A (Codec determines bitrate)

Audio Codec

YouTube/Vimeo (1080p HD): AAC

YouTube/Vimeo (4K UHD): AAC

High-Quality Archival Master: Uncompressed PCM (WAV)

Audio Bitrate

YouTube/Vimeo (1080p HD): 320 kbps

YouTube/Vimeo (4K UHD): 320 kbps

High-Quality Archival Master: 24-bit, 48 kHz

. . .

10.3 Preparing Your Film for the Festival Circuit

Submitting a documentary to film festivals involves adhering to a different set of technical standards than online streaming.[102] Most festivals now use online submission platforms like FilmFreeway, which allows for easy uploading of screener files.[104]

Submission Screener: For the initial submission and judging process, festivals typically require a high-quality online screener. A 1080p H.264 MP4 file, exported using the settings recommended for YouTube, is generally sufficient and widely accepted.[105] It is crucial to check the specific requirements of each festival, as they can vary.[102]

Exhibition Copy: If the film is accepted for screening at the festival, the organizers will request a high-resolution exhibition file. The requirements for this are much more stringent.

Digital File: Many festivals will request a high-quality master file, almost always in an **Apple ProRes 422 HQ** format, delivered as a.MOV file.[106] This file is much larger than a compressed H.264 file but maintains the high quality needed for projection.

Digital Cinema Package (DCP): For theatrical screenings in professional cinemas, the global standard is the DCP.[94] A DCP is a collection of files that contains the film's video, audio, and subtitles in a specific format designed for cinema projectors. While creating a DCP is a specialized process that may require third-party software or services, delivering a pristine ProRes master file is the necessary first step and the expected standard of professionalism for any filmmaker aspiring to the festival circuit.

11

THE LEGAL-EASE: DOCUMENTARY LAW & ETHICS

SHOOTING WITH A SMARTPHONE MAKES FILMMAKING MORE accessible and immediate than ever before. You can capture life as it happens, often without drawing the attention of a large camera crew. This power, however, comes with significant legal and ethical responsibilities. Ignoring them can, at best, prevent your film from ever being screened or, at worst, land you in a costly lawsuit.

This chapter is your legal primer. It is not a substitute for advice from a qualified entertainment lawyer, but it will equip you with the three core concepts you must understand before you ever press record: **Permission, Privacy, and Property.**

1. Permission: Getting Your Releases

A "release" is a legal contract that gives you permission to use a person's image or a location in your film. Without signed releases, you cannot sell, distribute, or even screen your documentary at most film festivals. Your smartphone's video app may not have a "sign here" button, but getting these forms signed is just as important as getting the shot.

Talent / Appearance Releases

This is the most critical form in your kit. It's a contract with your subject (everyone who is identifiable and speaks on camera) granting you the right to use their "likeness" (their face, voice, and name).

A proper talent release should include:

- **The Grant of Rights:** Language stating the person gives you the right to use their image, voice, and likeness in your film.
- **Media:** The right to use the footage in "any and all media, now known or hereafter devised." This future-proofs your film for new technologies.
- **Territory:** The right to distribute the film "throughout the world" or "in perpetuity" (forever).
- **Consideration:** A legal contract requires an exchange of value. This can be a nominal fee ($1.00 is common) or simply the "opportunity to appear" in the film.
- **For Minors:** If your subject is under 18, the release **must** be signed by their parent or legal guardian. A minor's signature is not legally binding.

Smartphone Pro-Tip: You don't need to carry a paper binder. Use an app like **Easy Release** or **Releases** (available on app stores) to get digital signatures on the spot. Or, keep PDF templates in your phone's "Files" app and use the "Markup" feature to have subjects sign with their finger.

Location Releases

If you film on private property, you need permission from the owner. This includes homes, offices, stores, and even a non-profit's headquarters.

- **Public vs. Private:** You generally **do not** need a release to film in a public place, like a city street, public park, or plaza. The exterior of a building, filmed from a public sidewalk, is also generally fine.
- **The "Gotcha":** "Publicly accessible" does not mean

"public property." A shopping mall, a coffee shop, or a grocery store is private property open to the public. The owners can (and often will) ask you to stop filming and leave. If you don't, you are trespassing. Always ask for the manager and get permission first.

2. Privacy: The Smartphone's Special Hurdle

Your phone's greatest asset—its discretion—is also its greatest legal liability, especially when it comes to audio.

The Reasonable Expectation of Privacy

The law protects people from being filmed in places where they have a "reasonable expectation of privacy." This includes obvious places like a bathroom, a private home (where you weren't invited), or a doctor's office. Using your phone to covertly film in these spaces is a serious violation.

On a public street, people do not have a reasonable expectation of privacy. You can film them. However, the law treats **audio** very differently from video.

The Big Trap: One-Party vs. Two-Party Consent

This is the most important legal concept for a smartphone filmmaker to understand. "Consent" refers to permission to *record audio* of a conversation.

- **One-Party Consent States:** In the majority of U.S. states (like New York and Texas), you only need the consent of **one person** in the conversation to record it. Since you are the filmmaker holding the phone, *you* are that one party. You can legally record your conversations with others without explicitly telling them.
- **Two-Party (or "All-Party") Consent States:** In states like **California, Florida, Pennsylvania, Illinois, and Washington** (among others), you must have the permission of *everyone* in the conversation to record it. If you secretly record a conversation in one of these states, you are committing a crime.

Always know the audio consent laws of the state you are in. When in doubt, always get verbal consent on camera. Simply starting an interview with, "Just so you know, this is being recorded for a documentary, is that okay with you?" can save you.

3. Property: Using Other People's Stuff (Copyright)

Your documentary will likely include music, news clips, or old photos. This content is someone else's intellectual property, and using it without permission is copyright infringement. You have two paths: **License It** or **Claim "Fair Use."**

Licensing Music & Footage

If you want to use a famous song or a clip from a movie, you must license it.

Music: You almost always need **two** licenses:

1 **Synchronization (Sync) License:** From the music publisher (the person who owns the written song/lyrics).

2 **Master Use License:** From the record label (the person who owns the *recording* of that song).

Archival/Stock Footage: You must get a license from the rights holder, whether it's a news organization (like AP, Reuters) or a stock footage site (like Getty, Shutterstock).

Licensing popular content is very expensive. For independent filmmakers, it's often better to use music from "royalty-free" or "Creative Commons" libraries, which grant you permission for a low flat fee or for free (with proper credit).

"Fair Use": The Filmmaker's Best Friend (and Biggest Risk)

Fair use is a legal doctrine that allows you to use small portions of copyrighted material *without permission* for purposes of criticism, commentary, or news reporting.

This is the legal right that allows a documentary to show

a clip from a news broadcast to analyze its coverage, or a clip from a movie to critique its message.

To determine if your use is "fair," courts apply a four-factor test:

1 **Purpose and Character:** Is your use "transformative"? (Are you using it for a new purpose, like criticism, rather than just replaying it?) This is the most important factor.

2 **Nature of the Work:** Using factual content (like a news clip) is more likely to be fair than using highly creative content (like a song or a scene from *Star Wars*).

3 **Amount Used:** Did you use only the small portion necessary to make your point, or did you use the "heart" of the work?

4 **Effect on the Market:** Does your use serve as a *substitute* for the original? (i.e., will people watch your clip *instead* of buying the original work?)

Warning: Fair use is not a *right*; it is an *affirmative defense*. This means if you are sued, you have to go to court and *prove* to a judge that your use was fair. This is a risk. Always consult the "**Documentary Filmmakers' Statement of Best Practices in Fair Use**" (linked below) as your guide.

4. Don't Forget Your Own Rights: Copyrighting Your Film

While the previous sections focused on respecting *other* people's rights, this one is about protecting *yours*.

It's a common misconception that you need to "apply for" a copyright. The truth is, **copyright protection is automatic the moment you create something.** The second you hit the stop-record button on your phone, you own the copyright to that footage. Your final edited documentary is a new, original work, and you own the copyright to that as well.

So, if it's automatic, why do people "copyright their

film"? What they are talking about is **registering** their copyright with the **U.S. Copyright Office**. This is a simple, low-cost administrative step that gives you powerful legal advantages.

Why Registration is Critical:

• **The Right to Sue:** In the United States, you **cannot file a lawsuit for copyright infringement** until your work is officially registered with the Copyright Office. Without registration, you have no legal recourse if someone steals your film.

• **Public Proof of Ownership:** Registration creates a public, searchable record that you are the legal owner of your film. This is essential for your "chain of title"—the legal documents that prove you own every part of your movie. Distributors, sales agents, and Errors & Omissions (E&O) insurance providers will demand this.

• **"Legal Firepower" (Statutory Damages):** This is the biggest financial reason. If you register your film *before* an infringement occurs (or within three months of its publication), you become eligible for **statutory damages and attorney's fees**. This means if you win your case, the court can award you a set amount of money (up to $150,000 for willful infringement) *plus* make the other side pay for your lawyers. Without this, you'd have to prove your exact financial losses, which is incredibly difficult for an independent film.

For a small fee (generally under $100), you can register your final film online at copyright.gov. This simple step transforms your automatic copyright from a shield into a sword, giving you the power to defend your work and sell it with confidence.

Do not just take our word for it. While these resources are essential reading for any documentary filmmaker, always consult with your attorney before undertaking any project to better understand the risks and how to protect yourself and your work.

Release Form Templates:

Talent Release: Smithsonian Folklife Sample Release Form (A great, simple template)

Location Release: Sample Location Release Form (A standard example)

Your Rights in Public:

ACLU: "Know Your Rights: Photographers"

Link: https://www.aclu.org/know-your-rights/photographers-rights

Why it's essential: This is the definitive guide on what you can and cannot film in public spaces, including encounters with police.

Audio Recording Laws:

Reporter's Committee for Freedom of the Press: "State-by-State Guide to Taping"

Link: https://www.rcfp.org/reporters-recording-guide/

Why it's essential: An interactive U.S. map that clearly explains the audio consent laws for every state. **Bookmark this.**

Copyright and Fair Use:

Documentary Filmmakers' Statement of Best Practices in Fair Use

Link: https://cmsimpact.org/code/documentary-filmmakers-statement-of-best-practices-in-fair-use/

Why it's essential: This is the foundational document, created by filmmakers and legal experts, that defines the community standard for fair use in documentaries.

U.S. Copyright Office: "More Information on Fair Use"

Link: https://www.copyright.gov/fair-use/more-info.html

Why it's essential: The official explanation of the four factors from the government office that registers copyright.

International Documentary Association (IDA)

Link: https://www.documentary.org/

Why it's essential: A vital professional organization with countless resources, articles, and guides on the legal and business side of documentary filmmaking.

12

THE ENDURING CORE IN THE EVER-PRESENT LENS

THE DEVICE IN YOUR POCKET, ONCE A SIMPLE TOOL FOR communication, has been reframed in your hands as a powerful instrument for documentary production. You have absorbed the technical knowledge—the nuances of exposure, the principles of sound capture, the art of the edit. But the true work of the documentarian begins now, moving beyond the craft of *how* to film and into the profound questions of *why* we film and *what* we seek to reveal.

The history of documentary is a testament to a fundamental human impulse: the desire to make sense of our world by telling true stories. This impulse is not bound by technology. As the great American documentarian Ken Burns reminds us, every filmmaker, regardless of their camera, is "under the power of the laws of storytelling". We are all compelled to find a beginning, a middle, and an end; to uncover conflict and seek resolution. While life rarely offers a tidy script, our task is to find what one storyteller calls the "story beneath the story that wants to be told". This structure is not an artificial imposition but the very grammar of human experience, and it is our primary responsibility to learn to recognize it in the beautiful chaos of reality.

That reality is best understood not through abstract data or sweeping statements, but through the intimate lens of individual lives. The most potent documentaries are not about issues; they are about people confronting issues. They invite us into a "core triangle" of filmmaker, subject, and audience, forging a bond of empathy that makes the story matter. Here, the smartphone grants you an extraordinary advantage. Its ubiquity and unassuming nature can dissolve the formal barrier that a large camera often erects. It fosters an intimacy that can, as was discovered in one of the earliest feature films shot on a mobile phone, lead people to "open up a little more easily". This proximity is a gift. It allows you to capture the subtle, authentic moments of vulnerability and transformation that are the lifeblood of a compelling character arc—the journey from who a person is to who they are becoming. Your duty is to honor that vulnerability, to portray the complex and often paradoxical nature of your subjects, resisting the temptation to reduce them to simple heroes or villains.

As you embark on this work, you will inevitably confront the central, haunting question of our field: what is truth? The masters offer no single map. Errol Morris, a titan of investigative documentary, stands firm in his conviction that "There is such a thing as truth," and that our work is a rigorous, obsessive pursuit of it, an attempt to figure out what really happened. Yet, the iconoclastic Werner Herzog challenges us to look beyond the surface, declaring that "Facts do not convey truth... truth creates illumination". He champions a "poetic, ecstatic truth," one that can only be reached "through fabrication and imagination and stylization" to touch a deeper human reality. The wise and playful

Agnès Varda saw documentary as an honest, subjective framing, admitting that "there is no one documentary that has all the pieces of the puzzle... you have to invent to understand to imagine the rest". And the observational master Frederick Wiseman pragmatically accepts that "Everything about a movie is manipulation," from the choice of subject to the final cut; the goal is not an impossible objectivity but a fair and honest interpretation.

You must find your own place within this philosophical spectrum. But one thing is non-negotiable: your ethical compass. The intimacy of your tool creates an "ethical proximity" that demands unwavering vigilance. The ease of capture must never lead to a poverty of care. Your primary responsibility is to prevent harm and to honor the trust of those who let you into their lives. As Barbara Kopple, who has spent her career immersed in the lives of her subjects, insists, the person being filmed must truly "want to be able to tell their story". This consent is not a piece of paper signed once, but a continuous, living dialogue.

Filmmaker	Core Philosophy on "Truth"	Key Quote
Werner Herzog	**Ecstatic Truth:** A deeper, poetic truth achieved through imagination and stylization, which illuminates more than mere facts.	"There are deeper strata of truth in cinema... reached only through fabrication and imagination and stylization."
Errol Morris	**Pursuit of Factual Truth:** A belief in an objective reality that can be investigated and approached, even if never fully grasped.	"There is such a thing as truth, but we often have a vested interest in ignoring it or outright denying it."
Agnès Varda	**Subjective Truth / Point of View:** Documentary is a collection of puzzle pieces, never the whole picture. It is an honest framing of a subjective reality.	"There is no one documentary that has all the pieces of the puzzle... there is no truth."
Frederick Wiseman	**Interpretive Truth:** All filmmaking is manipulation and choice. The goal is to create a fair interpretation of observed reality, not an objective record.	"Of course there's conscious manipulation! Everything about a movie is manipulation."

The smartphone has democratized the gaze, placing the power of production into the hands of anyone with a story to tell. It invites a new aesthetic, one born from its limitations. A shaky shot is not necessarily a mistake; it can be a visceral expression of a character's turmoil. Imperfect audio can signify the raw authenticity of a moment. The craft lies in knowing when to pursue polished cinematic language with gimbals and external mics, and when to embrace the raw, immediate, and haptic quality that is unique to this tool. As technologies like 5G and Artificial Intelligence further revolutionize our workflow , this core artistic and ethical judgment will only become more vital.

Ultimately, the greatest transformation this craft offers may not be in the films you create, but in the person you become. Frederick Wiseman once remarked, "Documentary filmmaking ruins you for real life, because you learn to be extremely attentive". This is the true gift. To wield the camera in your pocket is to commit to a practice of seeing. It trains you to notice the drama and grace in the ordinary, to listen more deeply, and to approach the world with a boundless curiosity. The technology will evolve, but the human need to bear witness, to connect, and to understand will not. The stories are all around you, waiting. You now have the tools. More importantly, you have begun to cultivate the vision. Go tell them.

AWARD-WINNING DOCUMENTARIANS

Here's a curated list of **award-winning documentarians** and **their most acclaimed works**, with reference links for further reading:

1. Errol Morris

Notable Works: *The Fog of War* (won Academy Award for Best Documentary Feature), *The Thin Blue Line* (critically acclaimed for helping overturn a wrongful conviction). (Wikipedia)

The Fog of War – https://www.imdb.com/title/tt0317910/

The Thin Blue Line – https://www.imdb.com/title/tt0096256/

More about him: https://en.wikipedia.org/wiki/Errol_Morris (Wikipedia)

2. Barbara Kopple

Notable Works: *Harlan County, USA* and *American Dream* — both won **Academy Awards for Best Documentary Feature.** (Wikipedia)

Harlan County, USA – https://www.imdb.com/title/tt0079945/

American Dream – https://www.imdb.com/title/tt0101476/

Bio & accolades: https://en.wikipedia.org/wiki/Barbara_Kopple (Wikipedia)

3. Michael Moore

Notable Works: *Bowling for Columbine, Fahrenheit 9/11* — both won significant awards including the **Academy Award (Bowling...)** and **Palm d'Or for Fahrenheit 9/11**. (Nree Productions)

Bowling for Columbine – https://www.imdb.com/title/tt0317648/

Fahrenheit 9/11 – https://www.imdb.com/title/tt0427944/

Profile: https://nreeproductions.com/best-documentary-filmmakers/ (Nree Productions)

4. Werner Herzog

Notable Works: *Grizzly Man, Cave of Forgotten Dreams* — acclaimed for immersive storytelling and exploratory documentaries. (Nree Productions)

Grizzly Man – https://www.imdb.com/title/tt0427327/

Cave of Forgotten Dreams – https://www.imdb.com/title/tt1093824/

Background: https://nreeproductions.com/best-documentary-filmmakers/ (Nree Productions)

5. Joshua Oppenheimer

Notable Works: *The Act of Killing, The Look of Silence* — both acclaimed for their bold treatment of historical atrocities. (Nree Productions)

The Act of Killing – https://www.imdb.com/title/tt2375605/

The Look of Silence – https://www.imdb.com/title/tt2119522/

Overview: https://nreeproductions.com/best-documentary-filmmakers/ (Nree Productions)

6. Ken Burns

Notable Works: *The Civil War*, *The Vietnam War* — celebrated for epic, deeply researched historical documentaries. (Nree Productions)

The Civil War – https://www.pbs.org/kenburns/the-civil-war/

The Vietnam War – https://www.pbs.org/show/vietnam-war/

Filmmaker info: https://nreeproductions.com/best-documentary-filmmakers/ (Nree Productions)

7. Eugene Jarecki

Notable Works: *Why We Fight*, *The King* — winner of Sundance Grand Jury Prizes, Golden Globe Documentary and major festival awards. (Wikipedia)

Why We Fight – https://www.imdb.com/title/tt0427944/

The King – https://www.imdb.com/title/tt6752376/

More on his career: https://en.wikipedia.org/wiki/Eugene_Jarecki (Wikipedia)

8. Rintu Thomas & Sushmit Ghosh

Most Known Work: *Writing With Fire* — won numerous festival awards and became the first Indian documentary nominated for the Academy Award for Best Documentary Feature. (Wikipedia)

Writing With Fire – https://www.imdb.com/title/tt14001646/

Read on their work:

- Rintu Thomas: https://en.wikipedia.org/wiki/Rintu_Thomas (Wikipedia)

• Sushmit Ghosh: https://en.wikipedia.org/wiki/Sushmit_Ghosh (Wikipedia)

9. Nicole Newnham

Notable Work: *Crip Camp* — Oscar-nominated and Emmy-winning documentary co-director/producer. (Wikipedia)

Crip Camp – https://www.imdb.com/title/tt8923486/

Bio: https://en.wikipedia.org/wiki/Nicole_Newnham (Wikipedia)

10. Chai Vasarhelyi & Jimmy Chin

Most Known Work: *Free Solo* — Oscar winner for Best Documentary and multiple Emmys. (people.com)

Free Solo – https://www.imdb.com/title/tt7775622/

Coverage: https://people.com/oscar-winning-free-solo-directors-chai-vasarhelyi-and-jimmy-chin-file-for-divorce-11724474 (people.com)

BIBLIOGRAPHY

1. No Camera? No Problem! Make a Great Documentary with Your Phone, https://beverlyboy.com/film-technology/no-camera-no-problem-make-a-great-documentary-with-your-phone/
2. Documentary mode, Wikipedia, https://en.wikipedia.org/wiki/Documentary_mode
3. Film Documentary Guide: 6 Types of Documentaries, 2025, MasterClass, https://www.masterclass.com/articles/film-documentary-guide
4. Nichols' 6 Modes of Documentary Might Expand Your Storytelling Strategies, https://nofilmschool.com/6-modes-of-documentary
5. Documentary genres. Criteria and dominant theories|International Conference on Cultural Informatics, Communication & Media Studies, eproceedings, https://eproceedings.epublishing.ekt.gr/index.php/cicms/article/view/2763
6. The 6 Types of Documentary Films, PremiumBeat, https://www.premiumbeat.com/blog/6-types-of-documentary-film/
7. UNIT 2 DIFFERENT MODES OF FILMMAKING*, eGyanKosh, https://www.egyankosh.ac.in/bitstream/123456789/72933/1/Unit-2.pdf
8. The six primary types of documentaries, Videomaker, https://www.videomaker.com/article/c06/18423-six-primary-styles-of-documentary-production/
9. Top tips for filming with a phone, ScreenSkills, https://www.screenskills.com/your-career/lesson-plans-and-classroom-resources/top-tips-for-filming-with-a-phone/
10. A Beginner's Guide to DJI Mic, Support, https://support.dji.com/help/content?customId=en-us03400007410&spaceId=34&re=US&lang=en&documentType=artical&paperDocType=paper
11. 4 Best Wireless Mics for Smartphones I've Tried and Trust,

Von Baron Music, https://vonbaronmusic.com/best-wireless-microphones-for-iphone-and-android/

12. Understanding Narrative Structure in Documentary Films, World Nomads, https://www.worldnomads.com/create/learn/film/understanding-narrative-structure-in-documentary
13. Master the Art of Documentary Storytelling in Three Acts, https://www.docfilmacademy.com/blog/master-the-art-of-documentary-storytelling-in-three-acts
14. Structuring a Documentary Narrative | Ken Burns Teaches Documentary Filmmaking, https://www.masterclass.com/classes/ken-burns-teaches-documentary-filmmaking/chapters/structuring-a-documentary-narrative
15. Ken Burns Shares 7 Tips for Structuring a Documentary, 2025, MasterClass, https://www.masterclass.com/articles/ken-burns-shares-tips-for-structuring-a-documentary
16. Documentary Structure | The Three-Act Documentary, Desktop-Documentaries.com, https://www.desktop-documentaries.com/documentary-structure.html
17. How To Craft A Documentary: choosing the right story structure for your film, https://www.actualityabroad.org/how-to-craft-a-documentary-film-story-structure/
18. Ultimate Guide to Phone Video, EditMate, https://www.editmate.com/ultimate-guide-to-phone-video/
19. How to Film a Documentary on Your Phone: A Beginner's Guide, https://www.docfilmacademy.com/blog/6-steps-to-getting-cinematic-footage-on-your-smartphone
20. Video Shooting in Manual Mode for Phones or Cameras, https://diyvideoeditor.com/video-shooting-in-manual-mode-for-phones-or-cameras/
21. Tips to film with mobile, Video, YouTube Help, https://support.google.com/youtube/answer/12948118?hl=en&co=YOUTUBE._YTVideoType%3Dvideo
22. Get the perfect photo with your smartphone camera's manual settings, Android Police, https://www.androidpolice.com/use-smartphone-camera-manual-settings-guide/
23. Use Manual Camera Controls like a Pro on Your Phone | Shutter Speed, ISO, WB, Exposure, https://www.youtube.com/watch?v=tuHnz8yBEno
24. Requesting beginner advice on smartphone/camera filming in various industries., Reddit, https://www.reddit.com/r/

Filmmakers/comments/1jex583/requesting_beginner_advice_on_smartphonecamera/

25. Best Practices and Tips for Shooting Smartphone Videos, Teaching Resources, https://teaching-resources.delta.ncsu.edu/best-practices-and-tips-for-shooting-smartphone-videos/
26. Support for DJI Mic 2, DJI, https://www.dji.com/support/product/mic-2
27. DJI Mic Mini | Using With a Phone, YouTube, https://www.youtube.com/watch?v=4TBD8quK7Wc
28. How to Connect a Wireless Lavalier Microphone? Detailed Steps with Examples, Hollyland, https://www.hollyland.com/blog/tips/connect-wireless-lavalier-microphone
29. How To Setup Wireless Lavalier Mic 2025: Only 5 Steps!, SYNCO, https://www.syncoaudio.com/blogs/news/how-to-setup-wireless-lavalier-mic
30. A Beginner's Guide to DJI Mic 2, Support, https://support.dji.com/help/content?customId=en-us03400009881&spaceId=34&re=US&lang=en&documentType=artical&paperDocType=paper
31. Real-Time Audio Monitoring on Smartphone #iphonehack #shorts #iphoneaudio ..., YouTube, https://www.youtube.com/shorts/p9ZwdgF1fI4
32. What Are The 5 Major Problems With Wireless Microphones? | Blog, Klover Products, https://www.kloverproducts.com/blog/what-are-the-5-major-problems-with-wireless-microphones
33. Easy Fix for Wireless Microphone Popping/Static Sound, Be a Wedding DJ, https://beaweddingdj.com/fix-wireless-microphone/
34. Common Wireless Mic Problems & How To Fix Them | Springtree Media, https://www.springtree.net/audio-visual-blog/is-this-thing-on-wireless-woes-and-how-to-avoid-them/
35. How to Resolve Wireless Interference and Wireless Problems, Sweetwater, https://www.sweetwater.com/sweetcare/articles/how-to-resolve-wireless-interference-and-wireless-problems/
36. Problems With Wireless Microphones and How to Fix Them : 9 Steps, Instructables, https://www.instructables.com/Problems-with-Wireless-Microphones-And-How-To-Fix-/
37. Why You Need a Smartphone Gimbal: Elevate Your Content Creation, Henry's Camera, https://blog.henrys.com/why-you-

need-a-smartphone-gimbal-elevate-your-content-creation/

38. Shop Portable Phone Gimbals for Stabilized Videos – DJI, https://www.dji.com/phone-gimbal
39. Gimbals, Insta360 Store, https://store.insta360.com/collections/gimbals
40. The best smartphone gimbals for iPhones and Android phones: make your phone footage silky smooth | Digital Camera World, https://www.digitalcameraworld.com/buying-guides/best-smartphone-gimbals-for-iphones
41. Mobile Videography Lighting: Setup Guide & Best Kits, TapeReal, https://web.tapereal.com/blog/mobile-videography-lighting-setup-guide-and-best-kits/
42. 5 professional lighting setups for interviews, Videomaker, https://www.videomaker.com/how-to/lighting/lighting-design/5-professional-lighting-setups-for-interviews/
43. 3-Point Lighting and More Interview Lighting Setup Tips, NBCU Academy, https://nbcuacademy.com/interview-lighting/
44. Master 3-Point Lighting for Interviews & Studio Setups, YouTube, https://www.youtube.com/watch?v=1Sb4g8Ln1dg
45. Documentary Editing Tips for Working with Lots of Footage, PremiumBeat, https://www.premiumbeat.com/blog/documentary-footage-editing-tips-tricks/
46. Essential Guide to Editing Documentaries, Avid, https://www.avid.com/resource-center/how-to-edit-documentary
47. Asset Management Tips for Filmmakers, Raindance Film Festival, https://raindance.org/asset-management-tips-for-filmmakers/
48. The Art Of Organization: Download The Perfect Folder Structure For Filmmakers Here!, https://noamkroll.com/the-art-of-organization-download-the-perfect-folder-structure-for-filmmakers-here/
49. The Only Post-Production Folder Structure You'll Ever Need, 2022 ..., https://thepostflow.com/post-production/take-your-efficiency-to-the-next-level-with-a-professional-post-production-folder-structure/
50. How To Set Up A Folder Structure – Stillmotion Blog, https://www.stillmotionblog.com/folder-structure/
51. Creating a file structure for a whole video department? Help. : r/editors, Reddit, https://www.reddit.com/r/editors/

comments/66i1eo/creating_a_file_structure_for_a_whole_video/

52. 5 Pro Tips for Organizing Your Video Files, Artgrid, https://artgrid.io/insights/how-to-organize-your-video-files/
53. Advice on tagging, organizing vast amount of footage (documentary), Creative COW, https://creativecow.net/forums/thread/advice-on-tagging-organizing-vast-amount-of-footag/
54. How to Edit a Documentary Film (Start To Finish), https://www.docfilmacademy.com/blog/how-to-edit-a-documentary-film
55. How to Use Digital Asset Management in Movie Production?, Pics.io, https://blog.pics.io/digital-asset-management-in-movie-production/
56. Making a Documentary Film – The Discipline of Organizing: 4th Professional Edition, https://berkeley.pressbooks.pub/tdo4p/chapter/making-a-documentary-film/
57. How do you organize many hours of footage? Documentary style, media management help request : r/editors, Reddit, https://www.reddit.com/r/editors/comments/wpeay1/how_do_you_organize_many_hours_of_footage/
58. How & Why to Transcribe a Documentary Interview | Rev, https://www.rev.com/blog/how-to-transcribe-a-documentary-interview
59. How To Edit An Interview For A Documentary, NYFA, https://www.nyfa.edu/student-resources/how-to-edit-an-interview-for-a-documentary/
60. What Is Documentary Transcription? Transcription Services For Documentary Filmmakers, Ditto Transcripts, https://www.dittotranscripts.com/blog/what-is-documentary-transcription-transcription-services-for-documentary-filmmakers/
61. Descript – Edit Video Like a Doc | AI Video & Podcast Editor, https://www.descript.com/
62. Video editing tips for documentary filmmakers, Simon Says AI, https://www.simonsaysai.com/blog/documentary-editing-tips
63. Filmmakers: Best transcription for editing & subtitles, Sonix, https://sonix.ai/filmmakers

64. First time editing a documentary. I need some workflow advice. : r/editors, Reddit, https://www.reddit.com/r/editors/comments/d2ky8r/first_time_editing_a_documentary_i_need_some/
65. best audio transcription software for documentary interviews? : r/Filmmakers, Reddit, https://www.reddit.com/r/Filmmakers/comments/1j5i95d/best_audio_transcription_software_for_documentary/
66. Documentary and Corporate Video Transcription, https://www.productiontranscripts.com/documentary-and-corporate-video-transcription/
67. How Transcripts Help Support Documentary Filmmakers, Verbit, https://verbit.ai/enterprise/how-transcripts-help-support-documentary-filmmakers/
68. What is the best documentary/interview workflow when working with transcriptions & premiere? : r/editors, Reddit, https://www.reddit.com/r/editors/comments/5hjdpz/what_is_the_best_documentaryinterview_workflow/
69. How to Create a "Paper Edit" for your Documentary, YouTube, https://www.youtube.com/watch?v=LcpLK7nWTwY
70. Elevating the Paper Edit for Documentary-style Filmmaking, Reduct.Video, https://reduct.video/blog/paper-edits-for-documentary-filmmakers/
71. Best free video editing software of 2025: Fully tested and free to use ..., https://www.techradar.com/best/free-video-editing-software
72. The Best Video Editing Software We've Tested | PCMag, https://www.pcmag.com/picks/the-best-video-editing-software
73. Video Editing Apps: Premiere Pro, After Effects, Adobe Express, https://www.adobe.com/creativecloud/video.html
74. Professional video editing software | Adobe Premiere Pro, https://www.adobe.com/products/premiere.html
75. Basic Editing Process, Digital Promise, https://digitalpromise.org/initiative/360-story-lab/360-production-guide/investigate/360-media/post-production/basic-editing-process/
76. Rough cut and fine cut | Narrative Documentary Production Class Notes, Fiveable, https://fiveable.me/narrative-documentary-production/unit-6/rough-cut-fine-cut/study-guide/1PNRxgZzhMsqbyKl

77. Rough cut: What is a rough cut in film and when is it edited? | Adobe, https://www.adobe.com/creativecloud/video/post-production/cuts-in-film/rough-cut.html
78. Rough Cut: Film Editing Basics for Video Creators, TourBox, https://www.tourboxtech.com/en/news/rough-cut.html
79. Rough cut, Wikipedia, https://en.wikipedia.org/wiki/Rough_cut
80. Stages of the Editing Process, Video Post-Production Workflow Guide | Frame.io, https://workflow.frame.io/guide/editing-stages
81. 1 Hour of Smartphone Filmmaking Tips for Beginners, YouTube, https://www.youtube.com/watch?v=ZaqAkxHR-DM
82. How To Color Grade Smartphone Videos like a PRO, YouTube, https://www.youtube.com/watch?v=LAc-vX5sR5c
83. How do you guys color grade your footage when using 3 or more different devices? I always find it hard to match (doing it completely manual). It's just continuous grind... : r/premiere, Reddit, https://www.reddit.com/r/premiere/comments/19al7jl/how_do_you_guys_color_grade_your_footage_when/
84. How to Edit a Documentary (Step-by-Step), YouTube, https://www.youtube.com/watch?v=osBI7aS-rcI
85. COLOR GRADE VIDEOS on your PHONE Like a PRO Secret app, YouTube, https://www.youtube.com/watch?v=NLeOlSZkZJc
86. Top 5 Video Color Correction Apps for Android & iPhone, Movavi, https://www.movavi.com/learning-portal/color-correction-app.html
87. FAST, EASY & FREE! Cinematic Color Grading in CapCut | Video Editing Tutorial, YouTube, https://www.youtube.com/watch?v=TMnBLZy3WFg
88. Fix Your FLAT Smartphone Videos (A-Z Color Grading), YouTube, https://www.youtube.com/watch?v=31Se7R-hYrg
89. Every video format, codec, and container explained, api.video, https://api.video/blog/product-updates/every-video-format-codec-and-container-explained/
90. The Popular Video Codes, Their Pros and Cons, and Related File Formats, Cloudinary, https://cloudinary.com/guides/video-formats/a-primer-on-video-codecs
91. Video Codecs: What They Are & the Best Formats for Streaming, Mux, https://www.mux.com/articles/video-codecs

92. Video Codecs, List of the best codecs and how they work, GetStream.io, https://getstream.io/glossary/video-codecs/
93. Best Video Codec for Streaming & Quality in 2025: Top Options Compared, Dacast, https://www.dacast.com/blog/best-video-codec/
94. Mobile Documentaries, Which is the best export format in 2022? : r/editors, Reddit, https://www.reddit.com/r/editors/comments/yrhjxc/mobile_documen taries_which_is_the_best_export/
95. What are your best practices for exporting feature films for festivals? : r/editors, Reddit, https://www.reddit.com/r/editors/comments/z0ia3d/what_are_your_best_practices_for_exporting/
96. The Ultimate Guide to Exporting Professional-Quality Videos, https://www.filmsupply.com/articles/export-professional-quality-videos/
97. What are the best video export recommendations? | Pic-Time Help Center, https://help.pic-time.com/en/articles/10029222-what-are-the-best-video-export-recommendations
98. Best export settings for Premiere Pro, Adobe, https://www.adobe.com/creativecloud/video/hub/guides/best-export-settings-for-premiere-pro.html
99. What audio format is best for movies?, Quora, https://www.quora.com/What-audio-format-is-best-for-movies
100. Top 10 Audio File Formats, Sage Audio, https://www.sageaudio.com/articles/top-10-audio-file-formats
101. voice123.com, https://voice123.com/blog/the-voice-over-industry/audio-file-formats/#:~:text=Evaluate%20audio%20quality%3A%20Lossless%20audio,quality%20narration%2C%20offering%20less%20compression.
102. Film Festival Submission Guide | The L.A. Film School, https://www.lafilm.edu/blog/film-festival-submission-guide/
103. Submit To Festival, sundance.org, https://www.sundance.org/festivals/sundance-film-festival/submit/
104. How It Works, FilmFreeway, https://filmfreeway.com/pages/how-it-works
105. 2021 Film Festival Submission Guides: Features, Shorts, Episodics & More, SXSW, https://www.sxsw.com/film/2020/2021-film-festival-submission-guidelines/

106. 2024 Technical Specifications for Short Film Submission, South Carolina ETV, https://www.scetv.org/sites/default/files/2024-08/NextLens%20Video%20Submission%20Guidelines_0.pdf
107. 2023 Technical Specifications For Festival Presentation + Content Delivery Instructions, Sundance Institute, https://www.sundance.org/wp-content/uploads/2022/11/Technical-Specifications-For-Festival-Presentation.pdf

ABOUT THE AUTHORS

Dänna Wilberg has been an award-winning short filmmaker for over two decades. As a screenwriter, producer and director, she has captured cinematic moments across many genres, from fiction to commercial video production. She is also the prolific author of the Grace Simms and Suzanne Cash series.

Terry Shepherd first fell in love with the visual arts in elementary school when he was one of several featured in an educational television production about storytelling. He taught television production while attending Michigan State University and has enjoyed a long career as a corporate executive, entrepreneur and author.

Made in the USA
Monee, IL
04 November 2026

67376809R00049